INVESTIGATIONS GUIDE

Plants and Animals

Full Option Science System™
Developed at the Lawrence Hall of Science, University of California, Berkeley
Published and Distributed by Delta Education

FOSS Lawrence Hall of Science Team
Larry Malone and Linda De Lucchi, FOSS Project Codirectors and Lead Developers
Kathy Long, FOSS Assessment Director; David Lippman, Program Manager; Carol Sevilla, Publications Design Coordinator; Susan Stanley, Illustrator; John Quick, Photographer
FOSS Curriculum Developers: Brian Campbell, Teri Lawson, Alan Gould, Susan Kaschner Jagoda, Ann Moriarty, Jessica Penchos, Kimi Hosoume, Virginia Reid, Joanna Snyder, Erica Beck Spencer, Joanna Totino, Diana Velez, Natalie Yakushiji
Susan Ketchner, Technology Project Manager
FOSS Technology Team: Dan Bluestein, Christopher Cianciarulo, Matthew Jacoby, Kate Jordan, Frank Kusiak, Nicole Medina, Jonathan Segal, Dave Stapley, Shan Tsai

Delta Education Team
Bonnie A. Piotrowski, Editorial Director, Elementary Science
Project Team: Jennifer Apt, Sandra Burke, Joann Hoy, Kristen Mahoney, Jennifer McKenna

Thank you to all FOSS Grades K-5 Trial Teachers
Heather Ballard, Wilson Elementary, Coppell, TX; Mirith Ballestas De Barroso, Treasure Forest Elementary, Houston, TX; Terra L. Barton, Harry McKillop Elementary, Melissa, TX; Rhonda Bernard, Frances E. Norton Elementary, Allen, TX; Theresa Bissonnette, East Millbrook Magnet Middle School, Raleigh, NC; Peter Blackstone, Hall Elementary School, Portland, ME; Tiffani Brisco, Seven Hills Elementary, Newark, TX; Darrow Brown, Lake Myra Elementary School, Wendell, NC; Heather Callaghan, Olive Chapel Elementary, Apex, NC; Katie Cannon, Las Colinas Elementary, Irving, TX; Elaine M. Cansler, Brassfield Road Elementary School, Raleigh, NC; Kristy Cash, Wilson Elementary, Coppell, TX; Monica Coles, Swift Creek Elementary School, Raleigh, NC; Shirley Conner, Ocean Avenue Elementary School, Portland, ME; Sally Connolly, Cape Elizabeth Middle School, Cape Elizabeth, ME; Melissa Cook-Airhart, Harry McKillop Elementary, Melissa, TX; Melissa Costa, Olive Chapel Elementary, Apex, NC; Hillary P. Croissant, Harry McKillop Elementary, Melissa, TX; Rene Custeau, Hall Elementary School, Portland, ME; Nancy Davis, Martha and Josh Morriss Mathematics and Engineering Elementary School, Texarkana, TX; Nancy Deveneau, Wilson Elementary, Coppell, TX; Karen Diaz, Las Colinas Elementary, Irving, TX; Marlana Dumas, Las Colinas Elementary, Irving, TX; Mary Evans, R.E. Good Elementary School, Carrollton, TX; Jacquelyn Farley, Moss Haven Elementary, Dallas, TX; Corinna Ferrier, Oak Forest Elementary, Humble, TX; Allison Fike, Wilson Elementary, Coppell, TX; Barbara Fugitt, Martha and Josh Morriss Mathematics and Engineering Elementary School, Texarkana, TX; Colleen Garvey, Farmington Woods Elementary, Cary, NC; Judy Geller, Bentley Elementary School, Oakland, CA; Erin Gibson, Las Colinas Elementary, Irving, TX; Kelli Gobel, Melissa Ridge Intermediate School, Melissa, TX; Dollie Green, Melissa Ridge Intermediate School, Melissa, TX; Brenda Lee Harrigan, Bentley Elementary School, Oakland, CA; Cori Harris, Samuel Beck Elementary, Trophy Club, TX; Kim Hayes, Martha and Josh Morriss Mathematics and Engineering Elementary School, Texarkana, TX; Staci Lynn Hester, Lacy Elementary School, Raleigh, NC; Amanda Hill, Las Colinas Elementary, Irving, TX; Margaret Hillman, Ocean Avenue Elementary School, Portland, ME; Cindy Holder, Oak Forest Elementary, Humble, TX; Sarah Huber, Hodge Road Elementary, Knightdale, NC; Susan Jacobs, Granger Elementary, Keller, TX; Carol Kellum, Wallace Elementary, Dallas, TX; Jennifer A. Kelly, Hall Elementary School, Portland, ME; Brittani Kern, Fox Road Elementary, Raleigh, NC; Jodi Lay, Lufkin Road Middle School, Apex, NC; Melissa Lourenco, Lake Myra Elementary School, Wendell, NC; Ana Martinez, RISD Academy, Dallas, TX; Shaheen Mavani, Las Colinas Elementary, Irving, TX; Mary Linley McClendon, Math Science Technology Magnet School, Richardson, TX; Adam McKay, Davis Drive Elementary, Cary, NC; Leslie Meadows, Lake Myra Elementary School, Wendell, NC; Anne Mechler, J. Erik Jonsson Community School, Dallas, TX; Anne Miller, J. Erik Jonsson Community School, Dallas, TX; Shirley Diann Miller, The Rice School, Houston, TX; Keri Minier, Las Colinas Elementary, Irving, TX; Stephanie Renee Nance, T.H. Rogers Elementary, Houston, TX; Cynthia Nilsen, Peaks Island School, Peaks Island, ME; Elizabeth Noble, Las Colinas Elementary, Irving, TX; Courtney Noonan, Shadow Oaks Elementary School, Houston, TX; Sarah Peden, Aversboro Elementary School, Garner, NC; Carrie Prince, School at St. George Place, Houston, TX; Marlaina Pritchard, Melissa Ridge Intermediate School, Melissa, TX; Alice Pujol, J. Erik Jonsson Community School, Dallas, TX; Claire Ramsbotham, Cape Elizabeth Middle School, Cape Elizabeth, ME; Paul Rendon, Bentley Elementary, Oakland, CA; Janette Ridley, W.H. Wilson Elementary School, Coppell, TX; Kristina (Crickett) Roberts, W.H. Wilson Elementary School, Coppell, TX; Heather Rogers, Wendell Creative Arts & Science Magnet Elementary School, Wendell, NC; Alissa Royal, Melissa Ridge Intermediate School, Melissa, TX; Megan Runion, Olive Chapel Elementary, Apex, NC; Christy Scheef, J. Erik Jonsson Community School, Dallas, TX; Samrawit Shawl, T.H. Rogers School, Houston, TX; Nicole Spivey, Lake Myra Elementary School, Wendell, NC; Ashley Stephenson, J. Erik Jonsson Community School, Dallas, TX; Jolanta Stern, Browning Elementary School, Houston, TX; Gale Stimson, Bentley Elementary, Oakland, CA; Ted Stoeckley, Hall Middle School, Larkspur, CA; Cathryn Sutton, Wilson Elementary, Coppell, TX; Camille Swander, Ocean Avenue Elementary School, Portland, ME; Brandi Swann, Westlawn Elementary School, Texarkana, TX; Robin Taylor, Arapaho Classical Magnet, Richardson, TX; Michael C. Thomas, Forest Lane Academy, Dallas, TX; Jomarga Thompkins, Lockhart Elementary, Houston, TX; Mary Timar, Madera Elementary, Lake Forest, CA; Helena Tongkeamha, White Rock Elementary, Dallas, TX; Linda Trampe, J. Erik Jonsson Community School, Dallas, TX; Charity VanHorn, Fred A. Olds Elementary, Raleigh, NC; Kathleen VanKeuren, Lufkin Road Middle School, Apex, NC; Valerie Vassar, Hall Elementary School, Portland, ME; Megan Veron, Westwood Elementary School, Houston, TX; Mary Margaret Waters, Frances E. Norton Elementary, Allen, TX; Stephanie Robledo Watson, Ridgecrest Elementary School, Houston, TX; Lisa Webb, Madisonville Intermediate, Madisonville, TX; Matt Whaley, Cape Elizabeth Middle School, Cape Elizabeth, ME; Nancy White, Canyon Creek Elementary, Austin, TX; Barbara Yurick, Oak Forest Elementary, Humble, TX; Linda Zittel, Mira Vista Elementary, Richmond, CA

Photo Credits: © DragoNika/Shutterstock (cover); © iStockphoto/Tobias Johanson; © Laurie Meyer; © John Quick

Published and Distributed by Delta Education, a member of the School Specialty Family
The FOSS program was developed in part with the support of the National Science Foundation grant nos. MDR-8751727 and MDR-9150097. However, any opinions, findings, conclusions, statements, and recommendations expressed herein are those of the authors and do not necessarily reflect the views of NSF. FOSSmap was developed in collaboration between the BEAR Center at UC Berkeley and FOSS at the Lawrence Hall of Science.

FOSS®, "Full Option Science System" and the FOSS logo are trademarks of the Regents of the University of California.

Copyright © 2019 by The Regents of the University of California

Standards cited herein from NGSS Lead States. 2013. *Next Generation Science Standards: For States, By States.* Washington, DC: The National Academies Press. Next Generation Science Standards is a registered trademark of Achieve. Neither Achieve nor the lead states and partners that developed the Next Generation Science Standards was involved in the production of, and does not endorse, this product.

All rights reserved. Any part of this work may not be reproduced or transmitted in any form or by any means, electronic or mechanical, including photocopying and recording, or by an information storage or retrieval system without prior written permission. For permission please write to: FOSS Project, Lawrence Hall of Science, University of California, Berkeley, CA 94720 or foss@berkeley.edu.

Plants and Animals
Investigations Guide, 1487588
978-1-62571-281-3
Printing 7 – 5/2018
Webcrafters, Madison, WI

INVESTIGATIONS GUIDE

Plants and Animals

TABLE OF CONTENTS

Overview	**1**
Framework and NGSS	**29**
Materials	**51**
Technology	**61**
Investigation 1: Grass and Grain Seeds	**71**
Part 1: Lawns	80
Part 2: Mowing the Lawn	96
Part 3: Wheat	103
Part 4: Variation in Plants and Animals	117
Investigation 2: Stems	**129**
Part 1: Rooting Stem Cuttings	138
Part 2: Spuds	146
Part 3: New Plants from Cuttings	155
Investigation 3: Terrariums	**165**
Part 1: Setting Up Terrariums	176
Part 2: Animals in the Terrarium	185
Part 3: Habitat Match	199
Part 4: Squirrel Behavior	207
Investigation 4: Growth and Change	**223**
Part 1: Planting Bulbs	232
Part 2: Planting Roots	239
Part 3: Plant and Animal Growth	248
Assessment	**261**

Welcome to FOSS® Next Generation™

Getting Started with FOSS Next Generation for Grades K–2

Whether you're new to hands-on science or a FOSS veteran, you'll be up and running in no time and ready to lead your students on a fantastic voyage through the wonders of the natural and designed world.

Watch our short video series or browse the next few pages to get started!

Getting Started with FOSS: Meet Your Module video

Scan here or visit deltaeducation.com/goFOSS

Three-Dimensional Active Science

It's time to experience the three dimensions of the NGSS—**disciplinary core ideas**, **crosscutting concepts**, and **science and engineering practices**. Engage in rich investigations that immerse your students in real-world applications of important scientific phenomena, supported by just-in-time teaching tips and strategies.

Getting Started with Your Equipment Kit

Meet Your FOSS Module!

Your FOSS module includes one or more large boxes, called drawers, and two smaller boxes for the Teacher Toolkit, student books, and other equipment. Each drawer has a label on the front listing its contents. Your packing list is always in Drawer 1.

Permanent Equipment

Your equipment kit includes enough permanent equipment for up to 8 groups (32 students). This equipment is classroom-tested and expected to last 7–10 years.

Consumable Equipment

Your kit also includes consumable materials for three class uses. Convenient refill kits provide materials for three additional uses and are available through Delta Education.

Easy Set-up and Clean-up!

FOSS Next Generation equipment drawers are packed by investigation to facilitate prep and to make packing up for the next use a snap!

Order Refills Online

deltaeducation.com/refillcenter

Drawer sections include:

- Unique materials needed for one investigation
- Common equipment used in multiple investigations
- Consumable materials—when it's empty you know it's time to refill!

Live Organisms

Some investigations require live organisms. Schools are encouraged to purchase these organisms from a local biological supply company to minimize both transit time and the impact of adverse weather on the health of the organisms.

If living material cards are purchased from Delta Education, they will be shipped separately in a green and white envelope. Keep these cards in a safe place until it's time to redeem them for the investigation.

Call Delta Education at 800-258-1302 at least three weeks before you need your organisms.

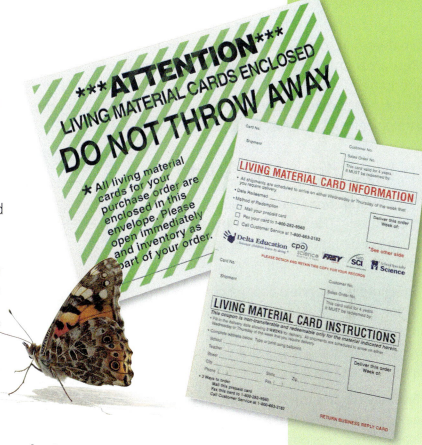

Premium Student eBook Access

If your school purchased a premium class license for the *FOSS Science Resources* student eBook, your access codes will be shipped separately in a blue and white striped envelope. Use this access code on FOSSweb to unlock student eBook access.

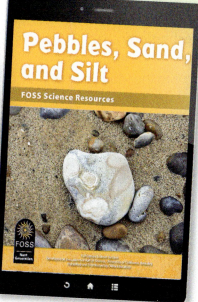

Getting Started with Your Teacher Toolkit

The Teacher Toolkit is the most important part of the FOSS program. There are three parts of the Teacher Toolkit—the **Investigations Guide**, **Teacher Resources**, and the student **Science Resources** book. It's here that all the wisdom and experience from years of research and classroom development comes together to support teachers with lesson facilitation and in-depth strategies for taking investigations to the next level.

1. Investigations Guide

The **Investigations Guide** is your roadmap to prepare for and lead the FOSS investigations. Chapters are tabbed for easy access to important module information.

The module **Overview** chapter gives you a high-level look at the 10–12 weeks of instruction in each module including a summary matrix, schedule for the module, and product support contacts.

Framework and the NGSS chapter provides a complete overview of NGSS connections, learning progressions, and background to support the conceptual framework for the module.

The **Materials** chapter is a must-read resource that helps you get your student equipment ready for first-time use and shares helpful tips for getting your classroom ready for FOSS.

The **Technology** chapter provides an overview for each digital resource in the module and gets you up and running on FOSSweb.com, complete with technical support.

Each **Investigation** chapter includes an At-a-Glance overview, science background content with NGSS connections, and in-depth guidance for preparing and facilitating instruction.

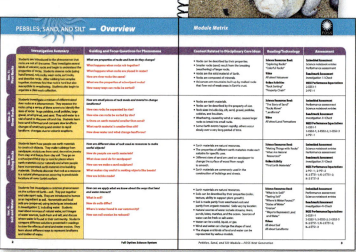

Module matrix

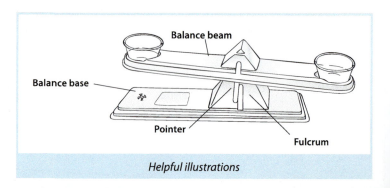

Helpful illustrations

The At-a-Glance chart includes:
- Summaries and pacing for investigation scheduling
- Focus questions for investigative phenomena
- Connections to disciplinary core ideas
- Reading, writing, and technology integration opportunities
- Embedded and benchmark assessments

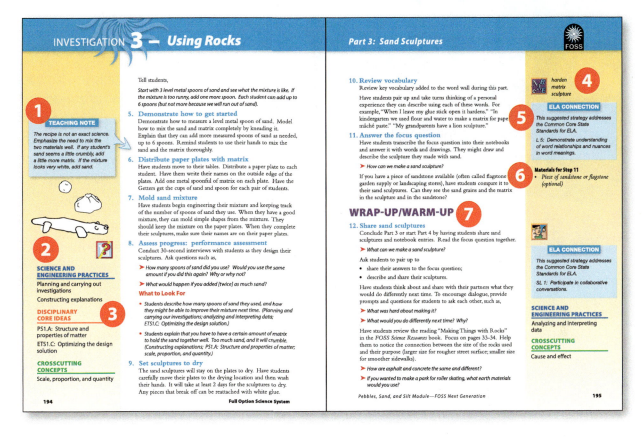

FOSS investigations provide the right support, when you need it with point-of-use guidance.

1. Teaching notes to facilitate instruction
2. Key three-dimensional highlights
3. Embedded assessment "What to Look For" in grades 1–2
4. Vocabulary review
5. Strategies to support English Language Arts
6. Materials used in the current steps
7. Discussion questions and model responses

The **Assessment** chapter gives you an in-depth look at the research-based components of the FOSS Assessment System, guidance on assessing for the NGSS, and generalized next-step strategies to use in your classroom. Find duplication masters, assessment charts, coding guides, and specific next-step strategies on FOSSweb.com.

Getting Started with Your Teacher Toolkit

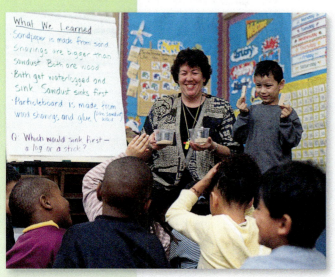

2. Teacher Resources

Your *Investigations Guide* tells you how to facilitate each investigation of a module. The **Teacher Resources** provides guidance on how to do it at your grade level across three modules throughout the year with effective practices and strategies derived from research and extensive field-testing.

A grade-level **Planning Guide** provides an overview to your three modules and an introduction to three-dimensional teaching and learning.

The **Science Notebooks** chapter provides age-appropriate methods to support students in developing productive science notebooks. Access powerful research-based next-step strategies to maximize the effectiveness of the notebook as a formative assessment tool.

Science-Centered Language Development is a collection of standards-aligned strategies to support and enhance literacy development in the context of science—reading, writing, speaking, listening, and vocabulary development.

In **Taking FOSS Outdoors**, find guidance for managing the space, time, and materials needed to provide authentic, real-world learning experiences in students' schoolyards and local communities.

Teacher Resources also includes:

- Grade-level connections to Common Core ELA and Math standards
- Module-specific notebook, teacher, and assessment blackline masters.

Check FOSSweb for the latest updates to chapters in *Teacher Resources*.

3. FOSS Science Resources Student Book

The Teacher Toolkit includes one copy of the student book. Reading is an integral part of science learning. Reading informational text critically and effectively is an important component of today's ELA standards. Once students have engaged with phenomena firsthand, they go more in-depth with articles in *FOSS Science Resources*.

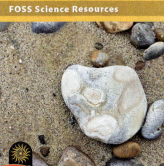

Module includes Big Book

Articles from FOSS *Science Resources* complement and enhance the active investigations, giving students opportunities to:
- Ask and answer questions
- Use evidence to support their ideas
- Use text to acquire information
- Draw information from multiple sources
- Interpret illustrations to build understanding

Interactive eBooks

FOSS Science Resources is available as a convenient, platform-neutral interactive student eBook with integrated audio, highlighted text, and links to videos and online activities. Student access to eBooks is available as an additional purchase.

Getting Started with Technology

FOSSweb.com

Easy access to program support resources

FOSSweb.com is your home for accessing the complete portfolio of digital resources in the FOSS program. Easily manage each of your modules, create class pages, and keep helpful references at your fingertips.

eInvestigations Guide

This easy-to-use interactive version of the *Investigations Guide* is mobile-friendly and offers simplified navigation, collapsible sections, and the ability to add customized notes.

Resources by Investigation

Easily access the duplication masters, online activities, and streaming videos needed for the current investigation part.

Teacher Preparation Videos

Videos provide helpful equipment setup instructions, safety information, and a summary of what students will do and learn throughout a part.

Interactive Whiteboard Lessons

Developed for SMART™ or Promethean boards, these resources help you facilitate each part of every investigation and give the class a visual reference.

Online Activities for Differentiating Instruction

FOSSweb digital resources provide engaging, interactive virtual investigations and tutorials that offer additional content and skill support for students. These experiences also help students who were absent catch up with class.

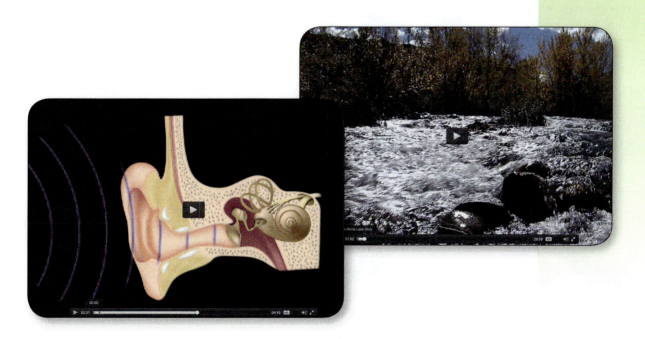

Streaming Videos

Videos are available on FOSSweb to support many investigations and often take students "on location" around the world or showcase experiments that would be too messy, expensive, or dangerous for the classroom.

Three-Dimensional Active Learning

The FOSS program has always placed student learning of science *practices* on equal footing with science *concepts and core ideas* and the NGSS and *Framework for K–12 Science Education* have provided a new language with which to articulate this. In each **FOSS Next Generation** investigation, students are engaged in the three dimensions of the NGSS to develop increasingly complex knowledge and understanding.

Science and engineering practices are the cognitive tools scientists and engineers use to answer questions and design solutions. FOSS students use these tools to gather evidence and to explain real-world phenomena.

Grade-level appropriate **disciplinary core ideas** are the concepts and established ideas of science. FOSS students develop these building blocks throughout investigations to make sense of phenomena.

Crosscutting concepts help students to connect the varied concepts and disciplines of science. FOSS students apply these concepts to different situations in order to make connections and develop comprehensive understanding.

FOSS Forward Thinking

The FOSS Vision

When the Full Option Science System (FOSS) began, the founders envisioned a science curriculum that was enjoyable, logical, and intuitive for teachers, and stimulating, provocative, and informative for students. Achieving this vision was informed by research in cognitive science, learning theory, and critical study of effective practice. The modular design of the FOSS product allowed users to select topics that aligned with district or state learning objectives, or simply resonated with their perception of comprehensive and reasonable science instruction. The original design of the FOSS Program was comprehensive in terms of coverage. FOSS was designed to provide real and meaningful student experience with important scientific ideas and to nurture developmentally appropriate knowledge of the objects, organisms, systems, and principles governing, the natural world.

The FOSS Next Generation Program

But the developers never envisioned FOSS to be a static curriculum, and now the Full Option Science System has evolved into a fully realized 21st century science program with authentic connection to the *Next Generation Science Standards (NGSS)*. The FOSS science curriculum is a comprehensive science program, featuring instructional guidance, student equipment, student reading materials, digital resources, and an embedded assessment system. The FOSS philosophy has always taken very seriously the teaching of good, comprehensive, accurate, science content using the methods of inquiry to advance that science knowledge. But the *Framework for K–12 Science Education*, on which the NGSS are based has allowed us to articulate our mission in a more coherent manner, using the vocabulary established by the authors of the *Framework*. The FOSS instructional design now strives to

a. communicate the disciplinary core ideas (content) of science, while

b. guiding and encouraging students to engage in or exercise the science and engineering practices (inquiry methods) to develop knowledge of the disciplinary core ideas, and

c. help students apprehend the crosscutting concepts (themes that unite core ideas, overarching concepts) that connect the learning experiences within a discipline and bridge meaningfully across disciplines as students gain more and more knowledge of the natural world.

> **The Full Option Science System has evolved into a fully realized 21st century science program with authentic connection to the Next Generation Science Standards (NGSS).**

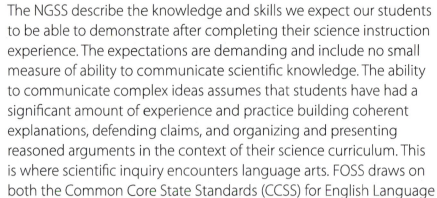

The NGSS describe the knowledge and skills we expect our students to be able to demonstrate after completing their science instruction experience. The expectations are demanding and include no small measure of ability to communicate scientific knowledge. The ability to communicate complex ideas assumes that students have had a significant amount of experience and practice building coherent explanations, defending claims, and organizing and presenting reasoned arguments in the context of their science curriculum. This is where scientific inquiry encounters language arts. FOSS draws on both the Common Core State Standards (CCSS) for English Language Arts and research data regarding the productive use of student science notebooks. FOSS developers realize that the most effective science program must seamlessly integrate science instruction goals and language arts skills. Science is one of the most engaging and productive arenas for introducing and exercising language arts skills: vocabulary, nonfiction (informational) reading, cause-and-effect relationships, on and on.

FOSS is strongly grounded in the realities of the classroom and the interests and experiences of the learners. The content in FOSS is teachable and learnable over multiple grade levels as students increase in their abilities to reason about and integrate complex ideas within and between disciplines.

FOSS is crafted with a structured, yet flexible, teaching philosophy that embraces the much-heralded 21st century skills; collaborative teamwork, critical thinking, and problem solving. The FOSS curriculum design promotes a classroom culture that allows both teachers and students to assume prominent roles in the management of the learning experience.

FOSS is built on the assumptions that understanding of core scientific knowledge and how science functions is essential for citizenship, that all teachers can teach science, and that all students can learn science. Formative assessment in FOSS creates a community of reflective practice. Teachers and students make up the community and establish norms of mutual support, trust, respect, and collaboration. The goal of the community is that everyone will demonstrate progress and will learn and grow.

PLANTS AND ANIMALS — *Overview*

Contents

Introduction 1
Module Matrix 2
FOSS Components 4
FOSS Instructional Design 8
Differentiated Instruction for Access and Equity 16
FOSS Investigation Organization 19
Establishing a Classroom Culture 22
Safety in the Classroom and Outdoors 25
Scheduling the Module 26
FOSS Contacts 28

The NGSS Performance Expectations addressed in this module include:

Life Sciences
1-LS1-1
1-LS1-2
1-LS3-1

Engineering, Technlogy, and Applications of Science
K–2 ETS1-2

▶ **NOTE**
The three modules for grade 1 in FOSS Next Generation are

Sound and Light

Air and Weather

Plants and Animals

INTRODUCTION

This module engages students with the anchor phenomenon that young plants and animals (offspring) have structures and behaviors that help them grow and survive. The driving question for the module is how do young plants and animals survive in their habitat? Students observe firsthand the structures of plants and discover ways to propagate new plants from mature plants (from seeds, bulbs, roots, and stem cuttings). They observe and describe changes that occur as young plants grow, and compare classroom plants to those in the schoolyard. They design terrariums (habitat systems) and provide for the needs of both plants and animals living together in the classroom.

Students explore the phenomenon of variation in the same kind of organism, including variation between young and adults. They learn about the behaviors of parents to help their young (offspring) survive. And they explore structure and function relationships as they sort different kinds of animal and plant structures. They use that understanding of structure and function, including animal sensory structures, to invent solutions to human problems.

Throughout the **Plants and Animals Module**, students engage in science and engineering practices by collecting and interpreting data to build explanations and designing and using tools to answer questions. Students gain experiences that will contribute to the understanding of the crosscutting concepts of patterns; cause and effect; systems and system models; and structure and function.

Full Option Science System

PLANTS AND ANIMALS — Overview

	Module Summary	Guiding and Focus Questions for Phenomena
Inv. 1: Grass and Grain Seeds	Students engage with the phenomeon of plant growth from seeds. They plant miniature lawns with ryegrass and alfalfa. They mow the lawns and observe the phenomenon of how grass and alfalfa respond to cutting. They plant individual wheat seeds in clear straws and observe early growth of plants, as well as variation in the growth of the same kind of seed. They conduct a schoolyard plant hunt and continue to look for variation. They use media to look at variation in kinds of animals and individuals of the same kind.	*What are the structures of a young plant growing from a seed?* What happens to ryegrass and alfalfa seeds in moist soil? What happens to the grass and alfalfa plants after we mow them? How does a wheat seed grow? How many different kinds of plants live in an area of the schoolyard?
Inv. 2: Stems	Students observe and describe the phenomenon of making new plants from stems of houseplants. They put sections of stems into water and look for evidence that a new plant is forming. Stem pieces that develop roots are planted to make new plants. Students plant pieces of potatoes (modified stems) and observe them grow.	*Where can new plants come from besides seeds?* How can we make a new plant from an old one? What grows from the nodes of a potato? How do we keep our cuttings alive?
Inv. 3: Terrariums	Students set up terrariums using seeds and plants from Investigations 1 and 2. They add local animals such as snails and isopods, provide for the needs of the organisms, and observe the phenomena of interactions. Students learn about other organisms through media and compare and sort structures and functions. Through an outdoor simulation, students engage with and describe the phenomena of variations in how squirrels store food for winter survival. Students engage with ways that engineers learn from nature to solve human problems.	*How do plants and animals survive in their habitat?* What do plants need to live and grow in a terrarium? What do animals need to live in a terrarium? What structures or behaviors do plants or animals have that help them live in their habitat? How do the behaviors of squirrels help them survive the winter?
Inv. 4: Growth and Change	Students plant bulbs in moist cotton and observe and describe the phenomenon of young plant development. They plant parts of roots—carrots and radishes—to discover which parts will develop into new plants and compare young to parent plants. Students adopt a schoolyard plant and compare it to other plants. They use media to learn about how behaviors of animals help their young to survive. Students describe the phenomenon of how young organisms resemble their parents.	*What do offspring get from their parents that help young survive?* How does a bulb grow? What parts of the parent plant can grow new plants? How do the plants in the schoolyard compare to the plants studied in class? What do animal parents do to help their young survive?

Module Matrix

Content Related to Disciplinary Core Ideas	Reading/Technology	Assessment
• Seeds need water to grow into new plants. • Not all plants grow alike. • Plant roots take in water and nutrients, and leaves make food from sunlight. • Seeds are alive and grow into new plants. • Plants have different structures that function in growth and survival. • Individuals of the same kind (of plant or animal) look similar but also vary in many ways.	**Science Resources Book** "What Do Plants Need?" "The Story of Wheat" "Variation" **Videos** *How Plants Grow* *Animal Growth*	**Embedded Assessment** Science notebook entries Performance assessment **Benchmark Assessment** *Investigation 1 I-Check* **NGSS Performance Expectations** 1-LS1-1; 1-LS3-1
• Leaves, twigs, and roots develop on stems at nodes. • Potatoes are underground stems; potato eyes are nodes where buds grow. • New plants can grow from the stems of mature plants. • Plants are living organisms that need, water, air, nutrients, light, and space to grow.		**Embedded Assessment** Science notebook entry Performance assessment **Benchmark Assessment** *Investigation 2 I-Check* **NGSS Performance Expectations** 1-LS1-1; 1-LS3-1
• Plants need water, nutrients, air, space, and light; animals need water, food, air, and space with shelter. • A habitat is a place where plants and animals live. It provides what a plant or animal needs to live. • Plants and animals live in different environments and have structures and behaviors that help them survive. Animals use sensory structures to take in information about their surroundings and act on it. • Engineers learn from nature to solve problems.	**Science Resources Book** "What Do Animals Need?" "Plants and Animals around the World" "Learning from Nature" **Videos** *How Plants Live in Different Places* *Animal Growth* **Online Activities** "Sorting Animals by Structures" "Habitat Sort"	**Embedded Assessment** Performance assessment Science notebook entries **Benchmark Assessment** *Investigation 3 I-Check* **NGSS Performance Expectations** 1-LS1-1; K–2 ETS1-2
• Plant bulbs are alive and grow new structures when provided with water. • Some parts of roots will grow into new plants if they are provided with water. Other parts will not. • Plants grow and change. Plants can produce new plants in many ways. • Adult animals can have young (offspring), and the young resemble their parents. • In many kinds of animals, parents and offspring engage in behaviors that help the young survive.	**Science Resources Book** "Animals and Their Young" **Video** *Animal Offspring and Caring for Animals* **Online Activities** "Watch It Grow!" "Find the Parent"	**Embedded Assessment** Science notebook entries Performance assessment **Benchmark Assessment** *Investigation 4 I-Check* **NGSS Performance Expectations** 1-LS1-2; 1-LS3-1

Plants and Animals Module—FOSS Next Generation

PLANTS AND ANIMALS — Overview

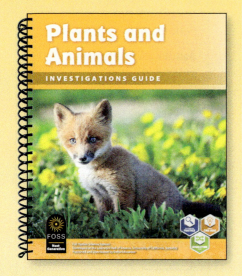

FOSS COMPONENTS

Teacher Toolkit for Each Module

The FOSS Next Generation Program has three modules for grade 1—Sound and Light, Air and Weather, and Plants and Animals.

Each module comes with a *Teacher Toolkit* for that module. The *Teacher Toolkit* is the most important part of the FOSS Program. It is here that all the wisdom and experience contributed by hundreds of educators has been assembled. Everything we know about the content of the module, how to teach the subject, and the resources that will assist the effort are presented here. Each toolkit has three parts.

Investigations Guide. This spiral-bound document contains these chapters.

- Overview
- Framework and NGSS
- Materials
- Technology
- Investigations (four in this module)
- Assessment

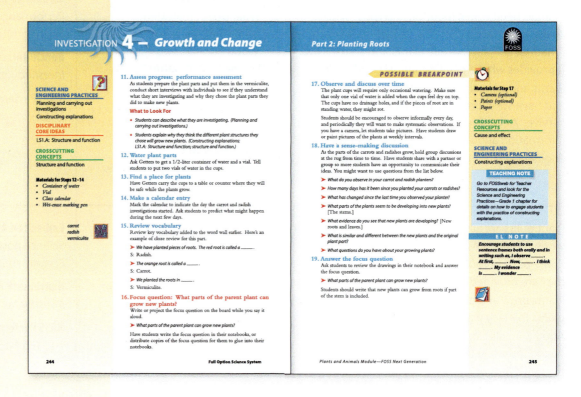

FOSS Components

FOSS Science Resources book. One copy of the student book of readings is included in the *Teacher Toolkit*.

Teacher Resources. These chapters can be downloaded from FOSSweb and are also in the bound *Teacher Resources* book.

- FOSS Program Goals
- Planning Guide—Grade 1
- Science and Engineering Practices—Grade 1
- Crosscutting Concepts—Grade 1
- Sense-Making Discussions for Three-Dimensional Learning—Grade 1
- Access and Equity
- Science Notebooks in Grades K–2
- Science-Centered Language Development
- FOSS and Common Core ELA—Grade 1
- FOSS and Common Core Math—Grade 1
- Taking FOSS Outdoors
- Science Notebook Masters
- Teacher Masters
- Assessment Masters

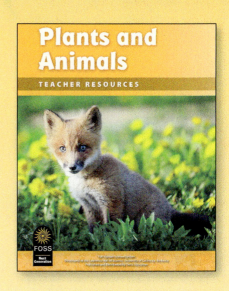

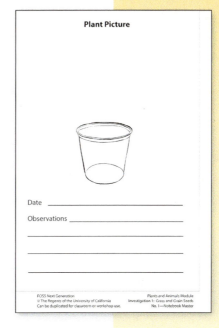

Equipment Kit for Each Module or Grade Level

The FOSS Program provides the materials needed for the investigations in sturdy, front-opening drawer-and-sleeve cabinets. Inside, you will find high-quality materials packaged for a class of 32 students. Consumable materials are supplied for three uses before you need to resupply. Teachers may be asked to supply small quantities of common classroom materials.

Delta Education can assist you with materials management strategies for schools, districts, and regional consortia.

Plants and Animals Module—FOSS Next Generation

PLANTS AND ANIMALS — *Overview*

> **NOTE**
> *FOSS Science Resources: Plants and Animals* is also provided as a big book in the equipment kit.

FOSS Science Resources Books

FOSS Science Resources: Plants and Animals is a book of original readings developed to accompany this module. The readings are referred to as articles in *Investigations Guide*. Students read the articles in the book as they progress through the module. The articles cover specific concepts, usually after the concepts have been introduced in the active investigation.

The articles in *FOSS Science Resources* and the discussion questions provided in *Investigations Guide* help students make connections to the science concepts introduced and explored during the active investigations. Concept development is most effective when students are allowed to experience organisms, objects, and phenomena firsthand before engaging the concepts in text. The text and illustrations help make connections between what students experience concretely and the ideas that explain their observations.

Some **offspring** get their first food from their parents. This young penguin is getting some seafood.
Can you see the camel getting milk from its mother?

As the young get older, some parents teach their offspring how to get food. Mother grizzly bears teach their cubs to catch fish.
The cub might have to try many times before catching a fish.

FOSS Components

Technology

The FOSS website opens new horizons for educators, students, and families, in the classroom or at home. Each module has digital resources for students and families—interactive simulations, virtual investigations, and online activities. For teachers, FOSSweb provides online teacher *Investigations Guides*; grade-level planning guides (with connections to ELA and math); materials management strategies; science teaching and professional development tools; contact information for the FOSS Program developers; and technical support. In addition FOSSweb provides digital access to PDF versions of the *Teacher Resources* component of the *Teacher Toolkit*, digital-only instructional resources that supplement the print and kit materials, and access to FOSSmap, the online assessment and reporting system for grades 3–8.

With an educator account, you can customize your homepage, set up easy access to the digital components of the modules you teach, and create class pages for your students with access to online activities.

▶ **NOTE**
To access all the teacher resources and to set up customized pages for using FOSS, log in to FOSSweb through an educator account. See the Technology chapter in this guide for more specifics.

Ongoing Professional Learning

The Lawrence Hall of Science and Delta Education strive to develop long-term partnerships with districts and teachers through thoughtful planning, effective implementation, and ongoing teacher support. FOSS has a strong network of consultants who have rich and experienced backgrounds in diverse educational settings using FOSS.

▶ **NOTE**
Look for professional development opportunities and online teaching resources on www.FOSSweb.com.

PLANTS AND ANIMALS — Overview

FOSS INSTRUCTIONAL DESIGN

FOSS is designed around active investigation that provides engagement with science concepts and science and engineering practices. Surrounding and supporting those firsthand investigations are a wide range of experiences that help build student understanding of core science concepts and deepen scientific habits of mind.

The Elements of the FOSS Instructional Design

Active Investigation

- Using Formative Assessment
- Integrating Science Notebooks
- Taking FOSS Outdoors
- Engaging in Science-Centered Language Development
- Accessing Technology
- Reading *FOSS Science Resources* Books

FOSS Instructional Design

Each FOSS investigation follows a similar design to provide multiple exposures to science concepts. The design includes these pedagogies.

- Active investigation in collaborative groups: firsthand experiences with phenomena in the natural and designed worlds
- Recording in science notebooks to answer a focus question dealing with the scientific phenomenon under investigation
- Reading informational text in *FOSS Science Resources* books
- Online activities to acquire data or information or to elaborate and extend the investigation
- Outdoor experiences to collect data from the local environment or to apply knowledge
- Assessment to monitor progress and inform student learning

In practice, these components are seamlessly integrated into a curriculum designed to maximize every student's opportunity to learn.

A **learning cycle** employs an instructional model based on a constructivist perspective that calls on students to be actively involved in their own learning. The model systematically describes both teacher and learner behaviors in a coherent approach to science instruction.

A popular model describes a sequence of five phases of intellectual involvement known as the 5Es: engage, explore, explain, elaborate, and evaluate. The body of foundational knowledge that informs contemporary learning-cycle thinking has been incorporated seamlessly and invisibly into the FOSS curriculum design.

Engagement with real-world **phenomena** is at the heart of FOSS. In every part of every investigation, the investigative phenomenon is referenced implicitly in the focus question that guides instruction and frames the intellectual work. The focus question is a prominent part of each lesson and is called out for the teacher and student. The investigation Background for the Teacher section is organized by focus question—the teacher has the opportunity to read and reflect on the phenomenon in each part in preparing for the lesson. Students record the focus question in their science notebooks, and after exploring the phenomenon thoroughly, explain their thinking in words and drawings.

In science, a phenomenon is a natural occurrence, circumstance, or structure that is perceptible by the senses—an observable reality. Scientific phenomena are not necessarily phenomenal (although they may be)—most of the time they are pretty mundane and well within the everyday experience. What FOSS does to enact an effective engagement with the NGSS is thoughtful selection of scientific phenomena for students to investigate.

▶ **NOTE**
The anchor phenomena establish the storyline for the module. The investigative phenomena guide each investigation part. Related examples of everyday phenomena are incorporated into the readings, videos, discussions, formative assessments, outdoor experiences, and extensions.

Plants and Animals Module—FOSS Next Generation

PLANTS AND ANIMALS — Overview

Active Investigation

Active investigation is a master pedagogy. Embedded within active learning are a number of pedagogical elements and practices that keep active investigation vigorous and productive. The enterprise of active investigation includes

- context: sharing prior knowledge, questioning, and planning;
- activity: doing and observing;
- data management: recording, organizing, and processing;
- analysis: discussing and writing explanations.

Context: sharing, questioning, and planning. Active investigation requires focus. The context of an inquiry can be established with a focus question about a phenomenon or challenge from you or, in some cases, from students. (What do animals need to live in a terrarium?) At other times, students are asked to plan a method for investigation. This might start with a teacher demonstration or presentation. Then you challenge students to plan an investigation, such as to find out what grows from the nodes of a potato. In either case, the field available for thought and interaction is limited. This clarification of context and purpose results in a more productive investigation.

Activity: doing and observing. In the practice of science, scientists put things together and take things apart, observe systems and interactions, and conduct experiments. This is the core of science—active, firsthand experience with objects, organisms, materials, and systems in the natural world. In FOSS, students engage in the same processes. Students often conduct investigations in collaborative groups of four, with each student taking a role to contribute to the effort.

The active investigations in FOSS are cohesive, and build on each other to lead students to a comprehensive understanding of concepts. Through investigations and readings, students gather meaningful data.

Data management: recording, organizing, and processing. Data accrue from observation, both direct (through the senses) and indirect (mediated by instrumentation). Data are the raw material from which scientific knowledge and meaning are synthesized. During and after work with materials, students record data in their science notebooks. Data recording is the first of several kinds of student writing.

Students then organize data so they will be easier to think about. Tables allow efficient comparison. Organizing data in a sequence (time) or series (size) can reveal patterns. Students process some data into graphs, providing visual display of numerical data. They also organize data and process them in the science notebook.

FOSS Instructional Design

Analysis: discussing and writing explanations. The most important part of an active investigation is extracting its meaning. This constructive process involves logic, discourse, and prior knowledge. Students share their explanations for phenomena, using evidence generated during the investigation to support their ideas. They conclude the active investigation by writing a summary of their learning in their science notebooks, as well as proposing answers to questions raised during the activity.

Science Notebooks

Research and best practice have led FOSS to place more emphasis on the student science notebook. Keeping a notebook helps students organize their observations and data, process their data, and maintain a record of their learning for future reference. The process of writing about their science experiences and communicating their thinking is a powerful learning device for students. The science-notebook entries stand as credible and useful expressions of learning. The artifacts in the notebooks form one of the core exhibitions of the assessment system.

You will find the duplication masters for grades 1–5 presented in notebook format. They are reduced in size (two copies to a standard sheet) for placement (glue or tape) into a bound composition book. Full-sized masters for grades 3–5 that can be filled in electronically and are suitable for display are available on FOSSweb. Student work is entered partly in spaces provided on the notebook sheets and partly on adjacent blank sheets in the composition book. Look to the chapter in *Teacher Resources* called Science Notebooks in Grades K–2 for more details on how to use notebooks with FOSS.

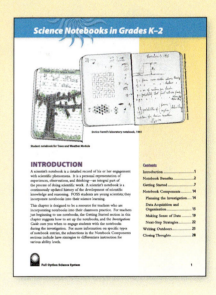

Plants and Animals Module—FOSS Next Generation

PLANTS AND ANIMALS — Overview

Reading in *FOSS Science Resources*

The *FOSS Science Resources* books, available in print and interactive eBooks, are primarily devoted to expository articles and biographical sketches. FOSS suggests that the reading be completed during language-arts time to connect to the Common Core State Standards for ELA. When language-arts skills and methods are embedded in content material that relates to the authentic experience students have had during the FOSS active learning sessions, students are interested, and they get more meaning from the text material.

Recommended strategies to engage students in reading, writing, speaking, and listening using the articles in the *FOSS Science Resources* books are included in the flow of Guiding the Investigation. In addition, a library of resources is described in the Science-Centered Language Development chapter in *Teacher Resources*.

The FOSS and Common Core ELA—Grade 1 chapter in *Teacher Resources* shows how FOSS provides opportunities to develop and exercise the Common Core ELA practices through science. A detailed table identifies these opportunities in the three FOSS modules for the first grade.

Engaging in Online Activities through FOSSweb

The simulations and online activities on FOSSweb are designed to support students' learning at specific times during instruction. Digital resources include streaming videos that can be viewed by the class or small groups. Resources can be used to review the active investigations and to support students who need more time with the concepts.

The Technology chapter provides details about the online activities for students and the tools and resources for teachers to support and enrich instruction. There are many ways for students to engage with the digital resources—in class as individuals, in small groups, or as a whole class, and at home with family and friends.

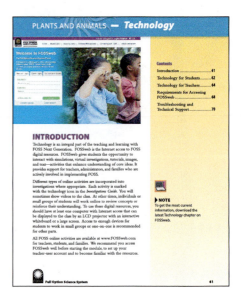

Full Option Science System

FOSS Instructional Design

Assessing Progress

The FOSS assessment system includes both formative and summative assessments. Formative assessment monitors learning during the process of instruction. It measures progress, provides information about learning, and is predominantly diagnostic. Summative assessment looks at the learning after instruction is completed, and it measures achievement.

Formative assessment in FOSS, called **embedded assessment**, is an integral part of instruction, and occurs on a daily basis. You observe action during class in a performance assessment or review notebooks after class. Performance assessments look at students' engagement in science and engineering practices or their recognition of crosscutting concepts, and are indicated with the second assessment icon. Embedded assessment provides continuous monitoring of students' learning and helps you make decisions about whether to review, extend, or move on to the next idea to be covered.

Benchmark assessments are short summative assessments given after each investigation. These **I-Checks** are actually hybrid tools: they provide summative information about students' achievement, and because they occur soon after teaching each investigation, they can be used diagnostically as well. Reviewing specific items on an I-Check with the class provides additional opportunities for students to clarify their thinking.

The embedded assessments are based on authentic work produced by students during the course of participating in the FOSS activities. Students do their science, and you observe the actions and look at their notebook entries. Bullet points in the Guiding the Investigation tell you specifically what students should know and be able to communicate.

If student work is incorrect or incomplete, you know that there has been a breakdown in the learning/communicating process. The assessment system then provides a menu of next-step strategies to resolve the situation. Embedded assessment is assessment *for* learning, not assessment *of* learning.

Assessment *of* learning is the domain of the benchmark assessments. Benchmark assessments for grades 1–2 are delivered after each investigation (I-Checks). These assessments can also be used to monitor and adjust instruction based on student understandings.

Plants and Animals Module—FOSS Next Generation

PLANTS AND ANIMALS — Overview

Taking FOSS Outdoors

FOSS throws open the classroom door and proclaims the entire school campus to be the science classroom. The true value of science knowledge is its usefulness in the real world and not just in the classroom. Taking regular excursions into the immediate outdoor environment has many benefits. First of all, it provides opportunities for students to apply things they learned in the classroom to novel situations. When students are able to transfer knowledge of scientific principles to natural systems, they experience a sense of accomplishment.

In addition to transfer and application, students can learn things outdoors that they are not able to learn indoors. The most important object of inquiry outdoors is the outdoors itself. To today's youth, the outdoors is something to pass through as quickly as possible to get to the next human-managed place. For many, engagement with the outdoors and natural systems must be intentional, at least at first. With repeated visits to familiar outdoor learning environments, students may first develop comfort in the outdoors, and then a desire to embrace and understand natural systems.

The last part of most investigations is an outdoor experience. Venturing out will require courage the first time or two you mount an outdoor expedition. It will confuse students as they struggle to find the right behavior that is a compromise between classroom rigor and diligence and the freedom of recreation. With persistence, you will reap rewards. You will be pleased to see students' comportment develop into proper field-study habits, and you might be amazed by the transformation of students with behavior issues in the classroom who become your insightful observers and leaders in the schoolyard environment.

▶ **NOTE**
The kit includes a set of four *Conservation* posters so you can discuss the importance of natural resources with students.

Teaching outdoors is the same as teaching indoors—except for the space. You need to manage the same four core elements of classroom teaching: time, space, materials, and students. Because of the different space, new management procedures are required. Students can get farther away. Materials have to be transported. The space has to be defined and honored. Time has to be budgeted for getting to, moving around in, and returning from the outdoor study site. All these and more issues and solutions are discussed in the Taking FOSS Outdoors chapter in *Teacher Resources*.

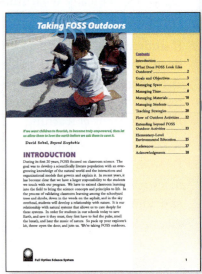

Full Option Science System

FOSS Instructional Design

Science-Centered Language Development and Common Core State Standards for ELA

The FOSS active investigations, science notebooks, *FOSS Science Resources* articles, and formative assessments provide rich contexts in which students develop and exercise thinking and communication. These elements are essential for effective instruction in both science and language arts—students experience the natural world in real and authentic ways and use language to inquire, process information, and communicate their thinking about scientific phenomena. FOSS refers to this development of language process and skills within the context of science as science-centered language development.

In the Science-Centered Language Development chapter in *Teacher Resources*, we explore the intersection of science and language and the implications for effective science teaching and language development. Language plays two crucial roles in science learning: (1) it facilitates the communication of conceptual and procedural knowledge, questions, and propositions, and (2) it mediates thinking—a process necessary for understanding. For students, language development is intimately involved in their learning about the natural world. Science provides a real and engaging context for developing literacy and language-arts skills identified in contemporary standards for English language arts.

The most effective integration depends on the type of investigation, the experience of students, the language skills and needs of students, and the language objectives that you deem important at the time. The Science-Centered Language Development chapter is a library of resources and strategies for you to use. The chapter describes how literacy strategies are integrated purposefully into the FOSS investigations, gives suggestions for additional literacy strategies that both enhance students' learning in science and develop or exercise English-language literacy skills, and develops science vocabulary with scaffolding strategies for supporting all learners. We identify effective practices in language-arts instruction that support science learning and examine how learning science content and engaging in science and engineering practices support language development.

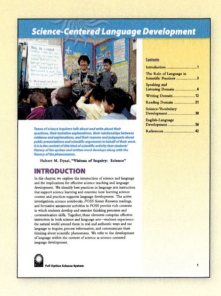

Specific methods to make connections to the Common Core State Standards for English Language Arts are included in the flow of Guiding the Investigation. These recommended methods are linked to the CCSS ELA through ELA Connection notes. In addition, the FOSS and the Common Core ELA chapter in *Teacher Resources* summarizes all of the connections to each standard at the given grade level.

Plants and Animals Module—FOSS Next Generation

PLANTS AND ANIMALS — *Overview*

DIFFERENTIATED INSTRUCTION FOR ACCESS AND EQUITY

Learning from Experience

The roots of FOSS extend back to the mid-1970s and the Science Activities for the Visually Impaired and Science Enrichment for Learners with Physical Handicaps projects (SAVI/SELPH Program). As this special-education science program expanded into fully integrated (mainstreamed) settings in the 1980s, hands-on science proved to be a powerful medium for bringing all students together. The subject matter is universally interesting, and the joy and satisfaction of discovery are shared by everyone. Active science by itself provides part of the solution to full inclusion and provides many opportunities at the same time for differentiated instruction.

Many years later, FOSS began a collaboration with educators and researchers at the Center for Applied Special Technology (CAST), where principles of Universal Design for Learning (UDL) had been developed and applied. FOSS continues to learn from our colleagues about ways to use new media and technologies to improve instruction. Here are the UDL guiding principles.

Principle 1. Provide multiple means of representation. Give learners various ways to acquire information and demonstrate knowledge.

Principle 2. Provide multiple means of action and expression. Offer students alternatives for communicating what they know.

Principle 3. Provide multiple means of engagement. Help learners get interested, be challenged, and stay motivated.

> *"Active science by itself provides part of the solution to full inclusion and provides many opportunities at the same time for differentiated instruction."*

Differentiated Instruction for Access and Equity

FOSS for All Students

The FOSS Program has been designed to maximize the science learning opportunities for all students, including those who have traditionally not had access to or have not benefited from equitable science experiences—students with special needs, ethnically diverse learners, English learners, students living in poverty, girls, and advanced and gifted learners. FOSS is rooted in a 30-year tradition of multisensory science education and informed by recent research on UDL and culturally and linguistically responsive teaching and learning. Procedures found effective with students with special needs and students who are learning English are incorporated into the materials and strategies used with all students during the initial instruction phase. In addition, the **Access and Equity** chapter in *Teacher Resources* (or go to FOSSweb to download this chapter) provides strategies and suggestions for enhancing the science and engineering experiences for each of the specific groups noted above.

Throughout the FOSS investigations, students experience multiple ways of interacting with phenomena and expressing their understanding through a variety of modalities. Each student has multiple opportunities to demonstrate his or her strengths and needs, thoughts, and aspirations.

The challenge is then to provide appropriate follow-up experiences or enhancements appropriate for each student. For some students, this might mean more time with the active investigations or online activities. For other students, it might mean more experience and/or scaffolds for developing models, building explanations, or engaging in argument from evidence.

For some students, it might mean making vocabulary and language structures more explicit through new concrete experiences or through reading to students. It may help them identify and understand relationships and connections through graphic organizers.

For other students, it might be designing individual projects or small-group investigations. It might be more opportunities for experiencing science outside the classroom in more natural, outdoor environments or defining problems and designing solutions in their communities.

Plants and Animals Module—FOSS Next Generation

PLANTS AND ANIMALS — Overview

English Learners

The FOSS Program provides a rich laboratory for language development for English learners. A variety of techniques are provided to make science concepts clear and concrete, including modeling, visuals, and active investigations in small groups. Instruction is guided and scaffolded through carefully designed lesson plans, and students are supported throughout.

Science vocabulary and language structures are introduced in authentic contexts while students engage in hands-on learning and collaborative discussion. Strategies for helping all students read, write, speak, and listen are described in the Science-Centered Language Development chapter. A specific section on English learners provides suggestions for both integrating English language development (ELD) approaches during the investigation and for developing designated (targeted and strategic) ELD-focused lessons that support science learning.

FOSS Investigation Organization

FOSS INVESTIGATION ORGANIZATION

Modules are subdivided into **investigations** (four in this module). Investigations are further subdivided into three to five **parts**. Each investigation has a general guiding question for the phenomenon students investigate, and each part of each investigation is driven by a specific **focus question**. The focus question, usually presented as the part begins, engages the student with the phenomenon and signals the challenge to be met, mystery to be solved, or principle to be uncovered. The focus question guides students' actions and thinking and makes the learning goal of each part explicit for teachers. Each part concludes with students recording an answer to the focus question in their notebooks.

The investigation is summarized for the teacher in the At-a-Glance chart at the beginning of each investigation.

Investigation-specific **scientific background** information for the teacher is presented in each investigation chapter organized by the focus questions.

The **Teaching Children about** section makes direct connections to the NGSS foundation boxes for the grade level—Disciplinary Core Ideas, Science and Engineering Practices, and Crosscutting Concepts. This information is later presented in color-coded sidebar notes to identify specific places in the flow of the investigation where connections to the three dimensions of science learning appear. The Teaching Children about section ends with information about teaching and learning and a conceptual-flow graphic of the content.

The **Materials** and **Getting Ready** sections provide scheduling information and detail exactly how to prepare the materials and resources for conducting the investigation.

Teaching notes and **ELA Connections** appear in blue boxes in the sidebars. These notes comprise a second voice in the curriculum—an educative element. The first (traditional) voice is the message you deliver to students. The second (educative) voice, shared as a teaching note, is designed to help you understand the science content and pedagogical rationale at work behind the instructional scene. ELA Connections boxes provide connections to the Common Core State Standards for English Language Arts.

FOCUS QUESTION
What do animal parents do to help their young survive?

SCIENCE AND ENGINEERING PRACTICES
Constructing explanations

DISCIPLINARY CORE IDEAS
LS1.B: Growth and development

CROSSCUTTING CONCEPTS
Patterns

TEACHING NOTE
This focus question can be answered with a simple yes or no, but the question has power when students support their answers with evidence. Their answers should take the form "Yes, because ____."

Plants and Animals Module—FOSS Next Generation

PLANTS AND ANIMALS — Overview

The **Getting Ready** and **Guiding the Investigation** sections have several features that are flagged in the sidebars. These include several icons to remind you when a particular pedagogical method is suggested, as well as concise bits of information in several categories.

The **safety** icon alerts you to potential safety issues related to chemicals, allergic reactions, and the use of safety goggles.

The small-group **discussion** icon asks you to pause while students discuss data or construct explanations in their groups.

The **new-word** icon alerts you to a new vocabulary word or phrase that should be introduced thoughtfully.

The **vocabulary** icon indicates where students should review recently introduced vocabulary.

The **recording** icon points out where students should make a science-notebook entry.

The **reading** icon signals when the class should read a specific article in the *FOSS Science Resources* book.

The **technology** icon signals when the class should use a digital resource on FOSSweb.

FOSS Investigation Organization

The **assessment** icons appear when there is an opportunity to assess student progress by using embedded or benchmark assessments. Some are performance assessments—observations of science and engineering practices, indicated by a second icon which includes a beaker and ruler.

The **outdoor** icon signals when to move the science learning experience into the schoolyard.

The **engineering** icon indicates opportunities for an experience incorporating engineering practices.

The **math** icon indicates an opportunity to engage in numerical data analysis and mathematics practice.

The **crosscutting concepts** icon indicates an opportunity to expand on the concept by going to *Teacher Resources*, Crosscutting Concepts chapter. This chapter provides details on how to engage students with that concept in the context of the investigation.

The **EL note** provides a specific strategy to use to assist English learners in developing science concepts.

EL NOTE

To help with pacing, you will see icons for **breakpoints**. Some breakpoints are essential, and others are optional.

POSSIBLE BREAKPOINT

Plants and Animals Module—FOSS Next Generation

PLANTS AND ANIMALS — Overview

ESTABLISHING A CLASSROOM CULTURE

Part of being a scientist is learning how to work collaboratively with others. However, students in primary grades are usually most comfortable working as individuals with materials. The abilities to share, take turns, and learn by contributing to a group goal are developing but are not as reliable as learning strategies all the time. Because of this egocentrism and the need for many students to control materials or dominate actions, the FOSS kit includes a lot of materials. To effectively manage students and materials, here are some suggestions.

Whole-Class Discussions

Introducing and wrapping up the center activities require you to work for brief periods with the whole class. FOSS suggests for these introductions and wrap-ups that you gather the class at the rug or other location in the classroom where students can sit comfortably in a large group.

At the beginning of the year, explain and discuss norms for sense-making discussions. You might start by together making a class poster with visuals to represent what it looks like, sounds like, and feels like when everyone is working and learning together. Model discussion protocols that give all students opportunities to speak and listen, such as think-pair-share. Review the norms before sense-making discussions, and leave time for reflecting on how well the group adhered to the norms. More strategies for developing oral discourse skills can be found in Sense-Making Discussions for Three-Dimensional Learning and the Science-Centered Language Development chapters in *Teacher Resources* on FOSSweb.

Collaborative Teaching and Learning

Collaborative learning requires a collective as well as individual growth mindset. A growth mindset is when people believe that their most basic abilities can be developed through dedication and hard work (see the research of Carol Dweck and her book *Mindset: The Psychology of Success*). As first-grade students learn to work together to make sense of phenomena and develop their inquiry and discourse skills, it's important to recognize and value their efforts to try new approaches, to share their ideas, and ask questions. Remind students that everyone in the classroom is a learner, and that learning happens when we try to figure things out. Here are a few ways to help students develop a growth mindset for science and engineering.

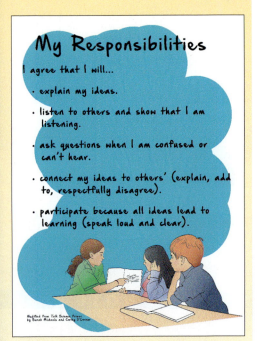

This poster is an example of student responsibilities that the class discussed and adopted as their norms.

Establishing a Classroom Culture

- **Praise effort, not right answers**. When students are successful at a task, provide positive feedback about their level of engagement and effort in the practices, e.g., the efforts they put into careful observations, how well they reported their observations, the relevancy of their questions, how well they connected or applied new concepts, and their use of new vocabulary, etc. Also, try to provide feedback that encourages students to continue to improve their learning and exploring, e.g., is there another way you could try? Have you thought about _____? Why do you think _____?

- **Foster and validate divergent thinking**. During sense-making discussions, continually emphasize how important it is to share emerging ideas and to be open to the ideas of others in order to build understanding. Model for students how you refine and revise your thinking based on new information. Make it clear to students that the point is not for them to show they have the right answer, but rather to help each other arrive at new understanding. Point out positive examples of students expressing and revising their ideas. For example, "Did you all notice how Carla changed her idea about _____?"

Establishing a classroom culture that supports three-dimensional teaching and learning centers on collaboration. Helping students to work together in pairs and small groups, and to adhere to norms for discussions, are ways to foster collaboration. These structures along with the expectations that students will be negotiating meaning together as a community of learners, creates a learning environment where students are compelled to work, think, and communicate like scientists and engineers to help one another learn.

Small-Group Centers

Some of the observations and investigations with earth materials can be conducted with small groups at a learning center. For example, students working with the plant and animal cards matching organisms to habitats in Investigation 3, Part 3, could be conducted at a center. Limit the number of students at the center to six to ten at one time. When possible, each student will have his or her own equipment to work with. In some cases, students will have to share materials and equipment and make observations together. As one group at a time is working at the center on a FOSS activity, other students will be doing something else. Over the course of an hour or more, plan to rotate all students through the center, or allow the center to be a free-choice station.

Plants and Animals Module—FOSS Next Generation

PLANTS AND ANIMALS — Overview

When You Don't Have Adult Helpers

Some parts of investigations work better when there is an aide or a student's family member available to assist groups with the activity and to encourage discussion and vocabulary development. We realize that there are many primary classrooms in which the teacher is the only adult present. You might invite upper-elementary students to visit your class to help with the activities. Remind older students to be guides and to let primary students do the activities themselves.

Managing Materials

The Materials section lists the items in the equipment kit and any teacher-supplied materials. It also describes things to do to prepare a new kit and how to check and prepare the kit for your classroom. Individual photos of each piece of FOSS equipment are available for printing from FOSSweb, and can help students and you identify each item. (Photo equipment cards are available in English and Spanish formats.)

For whole-class activities, FOSS Program designers suggest using a central materials distribution system. You organize all the materials for an investigation at a single location called the materials station. As the investigation progresses, one member of each group gets materials as they are needed, and another returns the materials when the investigation is complete. You place the equipment and resources at the station, and students do the rest. Students can also be involved in cleaning and organizing the materials at the end of a session.

When Students Are Absent

When a student is absent for a session, give him or her a chance to spend some time with the materials at a center. Another student might act as a peer tutor. Allow the student to bring home a *FOSS Science Resources* book to read with a family member. Each article has a few review items that the student can respond to verbally or in writing.

Safety in the Classroom and Outdoors

SAFETY IN THE CLASSROOM AND OUTDOORS

Following the procedures described in each investigation will make for a very safe experience in the classroom. You should also review your district safety guidelines and make sure that everything you do is consistent with those guidelines. Two posters are included in the kit: *FOSS Science Safety* for classroom use and *FOSS Outdoor Safety* for outdoor activities.

Look for the safety icon in the Getting Ready and Guiding the Investigation sections that will alert you to safety considerations throughout the module.

Safety Data Sheets (SDS) for materials used in the FOSS Program can be found on FOSSweb. If you have questions regarding any SDS, call Delta Education at 1-800-258-1302 (Monday–Friday, 8:00 a.m.–5:00 p.m. ET).

Science Safety in the Classroom

General classroom safety rules to share with students are listed here.

1. Listen carefully to your teacher's instructions. Follow all directions. Ask questions if you don't know what to do.
2. Tell your teacher if you have any allergies.
3. Never put any materials in your mouth. Do not taste anything unless your teacher tells you to do so.
4. Never smell any unknown material. If your teacher tells you to smell something, wave your hand over the material to bring the smell toward your nose.
5. Do not touch your face, mouth, ears, eyes, or nose while working with chemicals, plants, or animals.
6. Always protect your eyes. Wear safety goggles when necessary. Tell your teacher if you wear contact lenses.
7. Always wash your hands with soap and warm water after handling chemicals, plants, or animals.
8. Never mix any chemicals unless your teacher tells you to do so.
9. Report all spills, accidents, and injuries to your teacher.
10. Treat animals with respect, caution, and consideration.
11. Clean up your work space after each investigation.
12. Act responsibly during all science activities.

Plants and Animals Module—FOSS Next Generation

PLANTS AND ANIMALS — *Overview*

SCHEDULING THE MODULE

On the next page is a suggested teaching schedule for the module. The investigations are numbered and should be taught in order, as the concepts build upon each other. We suggest that 10 weeks be devoted to this module.

It is hard to keep organisms on a strict schedule. The general plan described below shows what the focus is each week. Many of the sessions are short observation sessions. Several plants will be investigated and growing concurrently.

The scheduled sessions involve **active investigation** as students observe plants growing, discuss changes in the plant parts, write in science notebooks, and learn new vocabulary in context. During **Wrap-Up/Warm-Ups** at the end of a part, students share notebook entries, and during **reading** sessions students read. *FOSS Science Resources* articles. Wrap-Up/Warm-Up and reading sessions can be completed during language-arts time to make connections to Common Core State Standards for ELA.

I-Checks are summative assessments at the end of each investigation.

Scheduling the Module

WEEK	1	2	3	4	5	6	7	8	9	10
1 Grass and Grain Seeds (14 sessions)	1.1 Lawns									
		1.2 Mowing Lawn	1.2 Mowing Lawn							
				1.3 Wheat	1.3 Wheat					
				1.4 Variation I-Check 1	1.4 Variation I-Check 1					
2 Stems (8 sessions)			2.1 Rooting Stem Cuttings	2.1 Rooting Stem Cuttings	2.1 Rooting Stem Cuttings	2.1 Rooting Stem Cuttings				
					2.2 Spuds	2.2 Spuds	2.2 Spuds			
						2.3 New Plants I-Check 2				
3 Terrariums (12 sessions)						3.1 Setting Up Terrariums				
						3.2 Animals in the Terrarium	3.2 Animals in the Terrarium			
							3.3 Habitat Match	3.3 Habitat Match		
								3.4 Squirrel Behavior I-Check 3		
4 Growth and Change (9 sessions)							4.1 Planting Bulbs	4.1 Planting Bulbs		
								4.2 Planting Roots	4.2 Planting Roots	
									4.3 Plant and Animal Growth I-Check 4	4.3 Plant and Animal Growth I-Check 4

> **NOTE**
> The Getting Ready section for each part of an investigation helps you prepare. It provides information on scheduling the activities and introduces the tools and techniques used in the activity. Be prepared—read the Getting Ready section thoroughly and review the teacher preparation video on FOSSweb.

PLANTS AND ANIMALS — *Overview*

FOSS CONTACTS

General FOSS Program information

www.FOSSweb.com

www.DeltaEducation.com/FOSS

Contact the developers at the Lawrence Hall of Science

foss@berkeley.edu

Customer Service at Delta Education

www.DeltaEducation.com/contact.aspx

Phone: 1-800-258-1302, 8:00 a.m.–5:00 p.m. ET

FOSSmap (online component of FOSS assessment system)

http://FOSSmap.com/

FOSSweb account questions/access codes/help logging in

techsupport.science@schoolspecialty.com

Phone: 1-800-258-1302, 8:00 a.m.–5:00 p.m. ET

School Specialty online support

loginhelp@schoolspecialty.com

Phone: 1-800-513-2465, 8:30 a.m.–6:00 p.m. ET

FOSSweb tech support

support@fossweb.com

Professional development

www.FOSSweb.com/Professional-Development

Safety issues

www.DeltaEducation.com/SDS

Phone: 1-800-258-1302, 8:00 a.m.–5:00 p.m. ET

For chemical emergencies, contact Chemtrec 24 hours a day.

Phone: 1-800-424-9300

Sales and replacement parts

www.DeltaEducation.com/FOSS/buy

Phone: 1-800-338-5270, 8:00 a.m.–5:00 p.m. ET